One day Paul saw a sign in the park. It said "Bike Race on Tuesday!"

"I'll enter that bike race," he said. "I'm the fastest rider I know."

Paul's bike was good, but it wasn't new. The rear wheel needed fixing. The tires needed checking. There were other things Paul would have to take care of before he could enter the race.

At home, Paul spread out his tools. First, he fixed the wheel so it turned smoothly. Then he patched a leaky tire. He got out some brushes and painted the bike dark blue.

"Now I'm ready for the race," said Paul.

On Tuesday, Paul coasted down Spring Street. There were many racers waiting at the starting line. Paul saw that Keith had a shiny new bike. It was black and had ten speeds. It looked bigger and faster than Paul's bike.

Keith proudly showed off his bike. "No one can beat me now," he boasted.

Paul did not feel hopeless. But he knew that Keith would be hard to beat. Quickly, Paul got in line with the other racers.

The race started with a bang. Keith led the way, with Paul not far behind. Paul rode fast, but Keith was faster. No matter what Paul did, he could not pass Keith. Keith's bike was the fastest of all. It looked like Keith would be the winner.

Suddenly Keith's bike swerved, and his brakes squealed. Paul saw Keith crash onto the road. Keith hadn't prejudged his speed. He hurt his knee, and his bike was wrecked.

Paul knew that he was the fastest rider now. He could win the race!

But it would be unkind to leave Keith. Paul had to think quickly. The other riders were nearer now. Should he keep going, or stop and help Keith?

Paul knew what he must do. He stopped his bike next to Keith. Paul was as helpful as could be.

Keith was surprised that Paul had stopped to help him, and he was thankful.

Paul and Keith finished the race last. Paul disliked being last, but he did not feel unhappy. He knew he had done the right thing. That made him feel like a winner. And he got a prize for being the most helpful racer of all! The judges have a fondness for racers who are good sports.